Keenly Waiting Online! Observation of Contemporary Interactive Culture and Technology-Based Society

JU-CHUN KO, Assistant Professor of
National Taipei University of Technology

Copyright © 2020 Dr. J. C. KO,
Assistant Professor of National Taipei University of Technology
All rights reserved.
ISBN: 9798570165830

DEDICATION

For my beloved Taipei, and Taipei Tech.

Keenly Waiting Online!
Observation of Contemporary Interactive Culture and Technology-Based Society

Radical Markets was selected as ***Books.com.tw*** Book of the Year 2020, and the Best Business and Economist Book of the Year by the Economist. The authors were both on Bloomberg's list of 50 Most Influential People. At the invitation of a publisher in Taiwan, the following three of the five chapters were especially written as extended reading hopefully to enlighten local readers in Taiwan. The other two chapters that have not been published are compiled here.

CONTENTS

	Acknowledgments	i
1	Preface: *Radical Markets* in Taiwan	1
2	Houses here look a bit old	4
3	Green card lottery meets diplomatic talents	9
4	Hand grappling techniques of the Monopolizing Octopus	15
5	Epilogue: Extended reading with local perspectives	20
6	One thousand ways of thinking for the 21st Century: *Reading 21 Lessons for the 21st century*, Y. N. Harari, author of *Sapiens* and *Homo Deus*	22
7	Follow Columbus in the Moments of History - A Foreword for Tim O'Reilly's new book *WTF? What's the Future and Why It's Up to Us*	27
8	"Can crowds be controlled?" *12 Steps: Face off with your own bran and your future self,*	32
9	Welcome to Zion; the fort of final truth requires our protection. Preface for *Deep Fakes and the Infocalypse: What You Urgently Need to Know*	36

ACKNOWLEDGMENTS

Department of Interaction Design,
National Taipei University of Technology.

1 PREFACE
RADICAL MARKETS IN TAIWAN

I recall a summer afternoon when I was in a renowned Taiwanese entrepreneur's office chatting over the immediate economic outlook that the world was about to face. It was a time before the deadly pneumonia outbreak, nor any need for home quarantine or WFH (working from home). What we saw was the revolutionary possibilities for the whole humanity brought by virtual reality, new media technologies, blockchain, Cryptocurrency and artificial intelligence. Looking outside the window, the entrepreneur told me about the income inequality that he witnessed recently. A while back, he went on a fund raising trip where he met all the rich business tycoons that any ordinary person may not get to meet in all three lifetimes. They were from Taiwan, China, Hong Kong, the US and Europe. He saw a three-story tall aquarium specifically for keeping two red dragon fish; a super sumptuous banquet where he downed a 2-millions-per-bottle red wine. As he talked, the entrepreneur rested his eyes on the distant mountains outside the window and said, "Look at the mountain there. Why doesn't it belong to us? Why are the mining rights of certain diamond mines and gold mines on the other side of the earth, despite being the property of all earth (citizens), inexplicably authorized by specific nations to private companies which would in turn impact on the global economy? Further, why are these assets and resources that should have been shared by all earth citizens be assigned, without asking, to some specific groups of people, be they nations, governments or consortia."

To a large extent, private properties turned public possessions somehow entail that you have a share in everything owned by other people around you.

Right. Governments are supposed to allocate resources on our behalf and benefit us. But why is it that what we see is governments taking our taxes to control banks by holding their shares, and the banks then loan out the money to consortia which use the money for further investments whose profits will ultimately return to the banks and governments, and on and on the transactions continue. Based on the concept of "Equality for all", the regular middle class should have been able to benefit from establishing social resources and maintaining political borders. But now it looks like only a small handful of people are always getting the profits. What is going wrong with the society?

This entrepreneur was known for his shrewd commentaries of current affairs, and many of his fans often had their innermost justice-related sentiments stirred up by the views that he published. Yet, the entrepreneur was an entrepreneur after all; he was no capitalist, so his opinions remained as opinions with little academic impact. When confronted, those in charge of power and resources could always nonchalantly claim that "Such are the rules, and I can only follow them," or "In a capitalist market, freedom reigns supreme." And that was that. Now that the book, *Radical Markets*, has been published, all the above would be given a new perspective. The book was co-written by Glen Weyl, the youngest ever Economics professor in the University of Chicago, and Eric Posner, the most distinguished member in the best-known family of legal scholars in the west. Vitalik Buterin, a tech genius in blockchain, instantly became a fan and volunteered to write a foreword for the book upon its publication and even gave a shout-out to the *Radical Markets* camp, "Whatever practical tech support you need, I am there," offering all his contacts and prestige in the tech circle and in the blockchain sector.

In *Radical Markets*, the two authors walk away from what is deemed by many as the alter of high achievers; like any social scientists that care about the current state of the world, they resort to mathematics, economics, social science and legal studies, as well as their understanding and imagination for technologies in an attempt to raise a warning against the contemporary capitalist economy that is walking on a tight rope. No, it's more than a warning. The two authors have shown their absolute courage in going against the tide to tell the truth and propose five fundamental remedies. They are like Gandalf and Frodo in our contemporary society where monopoly and totalitarianism are being carried out in the name of market competition.

As the social economics has developed in the past few decades, books like The Wealth of Nations by Adam Smith, A Treatise on Money by Keynes and Hayek's views on free market have all become the Rings of Power that entitle the ring owners, i.e. the capitalism practitioners, to speak with authority on topics like markets, capital and freedom. In reality, the capitalists have been carrying out one after another totalitarian scheme. Their so-called "Invisible

Hands" that are meant to be non-human hands but hands of divine intervention, and supposed to be part of the free market mechanism, have now become human interference playing God, trying to be invisible in their unjust dealings.

"The rich...are led by an invisible hand to make nearly the same distribution of the necessaries of life, which would have been made, had the earth been divided into equal portions among all its inhabitants, and thus without intending it, without knowing it, advance the interest of the society."

"Data, one of the most valuable commodities in the digital economy, are collected and monetized by companies such as Google and Facebook, but the users who create these data receive no direct compensation. A much-needed market in data simply does not exist."

The two authors of *Radical Markets* risk being misunderstood and scorned by the masses to unveil the reality behind the Rings. How we can follow them across high mountains to find the solution that may or may not exist depends on whether or not we take action after reading the book.

"Even if we don't sell you on all our ideas, we hope this book will open your mind to a new way of imagining the economy and politics. This challenging moment, when long-held assumptions are being overturned, is ripe for radical rethinking."

I was invited by the publisher of the book's traditional Chinese translation edition to write a response to Chapter 1, 3 and 4 from a local perspective to serve as extended reading for readers in Taiwan and those who can read traditional Chinese. The English translation of this book is intended for more readers to know Taiwan's perspectives on some of the innovative ideas in the book and their connection with Taiwan.

2 HOUSE HERE LOOK A BIT OLD

I recall a few years back when I was asked by a friend to host his friends from Shanghai on their first visit to Taiwan. We agreed to meet near the MRT Dongmen Station. We went through the alleyways. I as the host showed them the famous landmarks on Yong Kang Street and planned on taking them to sample soup dumplings in Din Tai Fung. All along the way, they showed great interest in the street scenes. I noticed one person was particularly curious about the buildings on both sides of Yong Kang Street. She was on the verge of saying something but then held back her words. I gathered whatever was on her mind might remain private or unresolved for a bit, so I went on to talk about the history of Yong Kong Street. I began with the Qing Dynasty when the area was a wilderness, went onto the construction of the Liugongjun Canal, the Bao Gong Cinema and the opening of the Da An Forest Park. This visitor from Shanghai finally broke her silence, looked at a row of under-five-story old houses on Yong Kang Street and said in her perfect Shanghai accent, "**How come your houses here all look a bit old?**"

It happened to be a time when the Wenlin Yuan Urban Renewal Dispute occurred and housing justice was being widely discussed; my friends with academic studies on public administration were also commenting on the various regulations of urban renewal. This visitor is a friend of a friend who went on my podcast as a guest before. On hearing her words, "Houses here look a bit old," I was speechless, not able to respond. I held my thoughts on quasi-democratic values and the price of totalitarianism, without uttering a word.

Notwithstanding this Shanghai visitor's words carried some undercurrents of misconception and even a bit of malicious mockery and attack. Now when I pass through certain streets that look distinctly out of place among its neighborhood, for instance, several sections on the Civic Boulevard or the corners around Qiang Shu High School behind the MRT Guting Station or Wenshan District and Nangang District as pointed out by many people on the social media, I can't help but think: wouldn't it be fantastic if there were an effective approach to urban renewal that could also protect the rights and interests of existing residents?

A few years on, I've become an assistant professor in National Taipei University of Technology (Taipei Tech). I am passing through Renai Road more often than ever where most of its sections have been turned into rich neighborhoods with luxurious houses. Yet, I came to notice one of the houses. Despite its ostensibly grand, beautiful and lavish exterior, the house was always locked up with leaves and cardboard boxes scattered all over the place. I could not possibly imagine how an opulent building like this on Renai Road would remain uninhabited. Had there been a case of Rose n' Siren Eyes or Taiwanese Paranormal Encounters occurring inside the house? I noted the address: No. 72, Sec. 2, Renai Road, and looked it up on internet to discover that it was the legendary luxury home which its housing developer built 16 years ago but was unable to sell. There were (and still are) so many people out there with no place to go to on a cold miserable night while this being one of the first grand houses on Renai Road had somehow remained unoccupied for 16 years. And that was only one of such examples. If you ever visit Taipei's Xinyi District, you will surely see this famous house standing as part of the Tao Zhu Yin Yuan luxury apartments dubbed "Spinning Luxury Homes". After going on the market for a few months, its first transaction was reportedly made with a subsidiary under the same construction company, hence the widespread speculation that the house was sold as an insider trading.

Evidently, monopolizing assets and leaving them idle are a waste to the entire society. The proposal of Luxury Tax, Empty Houses Tax and Housing Tax for Over-Accumulation are applauded by many people, while some others find it unnecessary and that it will interfere with people's rights to freely possess and control assets in a capitalist society, hence such mentality as the following: "What is wrong with me wanting to leave the house unoccupied after buying it?" or "What's wrong with wanting to buy an old metal house in the capital and live in it for 100 years?"

From the perspectives of *Radical Markets*, resource misallocation and immobility has to be resolved. With rational analysis, Glen Weyl as an economist states clearly that many ineffectually incurred taxes that are intended to solve all kinds of injustice are all missing the mark, and more strictly speaking, an unnecessary waste.

Is there a kind of tax system which, once announced, will save all the time spent on tax discussion and solve these problems in one go: resource idleness & waste and a small handful of people monopolizing the assets and refusing to comply with the bigger need for social environments and development and undertake adjustments accordingly? Housing justice is only one aspect, others including transportation development (with the case of Hyperloop, a high-speed metropolitan network, being cited by the authors as an example). Is there a tax system that can solve all the problems?

Indeed, the authors propose such a Hollywood-grade innovative tax system called the Common Ownership Self-Assessed Tax (COST). The so-called self-assessment means you can determine the value of an item yourself and pay the tax accordingly. If you consider your house worthy of $1, then you pay one dollar's worth of asset tax. Or you can define your house as worthy as $10 billion (so that you appear to be rich) and will by contrast pay the amount of tax corresponding to the $10 billion asset (the book citing 7% as the ideal COST tax, so that amounts to $0.7 billion in tax per year). The common ownership means that anyone can go through the auction mechanism (supplemented by some necessary resources/assets) as a protection mechanism to auction in (acquire) the resources/assets that you have put a price on. That is to say, you have lost the right to perpetual ownership of most commodities, and all the resources and assets that have an impact on the social development can be traded anytime and remain fluid, as is secured by compulsory law enforcement.

On reading this, many people, I believe, would exclaim: Isn't this communism? Isn't this totalitarianism?

Relax. Yes, to some degree, the authors' COST proposal is indeed made in the spirit of Communism or Socialism, but essentially, what lies behind the proposal is not a farther-reaching edition of Totalitarianism but an integration of extreme free market transactions in a bid to resolve many of the aforementioned resource idleness/misallocation impasse. But is their viewpoint really feasible, i.e. "no longer owning most of the things". Many criticize the authors' proposal as Socialist or Communist. But truth be told, we do live in a time when we can accept AirBnB as a co-shared housing solution, and when Japan has launched a new housing approach, i.e. HafH, offering "Fixed price all-you-can-stay" service, whereby monthly payments of a certain amount will give you unlimited access to many hundreds of studios and houses around the world. Similar trends include co-shared spaces and scooters, for instance, WEMO and GoShare in Taiwan. If we can accept having no permanent ownership of any item, then why can't we extend such concept/design of asset ownership to cover almost all items?

The authors reference the god of auction, Nobel Laureate in economics, William Vickrey, and call such assets covered by common ownership as Vickrey Commons. Mind you, this is not about co-sharing properties, but about being able to auction off the ownership of common properties at any time. To a larger extent, the ownership can be changed to a common entitlement. Alternatively, it's not a common ownership but a mechanism where its ownership can be possessed, governed, entitled and traded at any time. Such an extreme market viewpoint can be applied to the shared ownership of human resources, i.e., the COST of human capital.

"The talented enjoy a kind of freedom, as they can select from among a variety of appealing jobs. These jobs allow them to quickly accumulate capital that they can depend on, as they age, if they do not like the jobs that are available, or pick and choose among different levels of labor (part-time, enjoyable or low-paying jobs in the nonprofit sector, etc.). Those with fewer marketable skills are given a stark choice: undergo harsh labor conditions for low pay, starve, or submit to the many indignities of life on welfare. Yet the waste of social resources when a talented person fails to realize her potential is far greater, and arguably their failure to work should be punished more harshly.

A COST on human capital would ameliorate this form of unequal freedom by requiring the talented people to pay a tax if they do not want to work in a job that is most efficient for society."

To me, the above is one of the most shocking paragraphs in the book.

Are we able to accept such radical ideas? I often hear my friends talking about at what age they would like to start an early retirement. They make it sound as if they work hard enough, save enough capital by a certain age, they will be entitled to a life of no more work but play. Yet, the *Radical Markets* authors being the high-achievers stand with us on this matter in the sense that they resort to their economics expertise and propose that the injustice of such thinking, i.e. early retirement upon having accumulated enough personal wealth, desperately needs to be brought to our attention and debated over. After reading Chapter 1, I happened upon a translated article on BBC Chinese website published in 2018, A British Woman's Confession: Poverty opened my eyes to money. Excerpted from the article is as below:

"In the university dormitory, no matter where we came from, we all look more or less the same from the exterior. But during holidays, these new friends of mine would all disappear. They have all either gone home or taken on internship in London. Internship sounds great but it's not a paying position. I have no money to go on (non-paying) internship, so I worked in a local shoe shop selling shoes that I myself cannot afford. In so doing, I can pay for my rent and carry on living in the university dormitory... In our first class meeting, I felt what I knew fell far behind the guy that had done 10 intern jobs."

This instantly reminds me of my friends living in luxury homes who went to study in New York during school holidays while they were still in senior high school or university; they took up part-time jobs with McKinsey, while I worked as an intern for NT$50 a day producing online advertisements in a company located in a residential building in the Zhonghe Industrial Zone. What's more startling is that originally I was only envious/jealous of my friends living in luxury homes, but little did I know that being competent enough to be earn a wage of NT%50 a day was already deemed a triumph. There were people that could not get intern jobs and thus lost the opportunity to surpass or get equal with their peers. Based on this recollection and reflection, plus the following astounding data quoted by the authors:

"...we see that the share of income taken by the top 1% of earners has roughly doubled from its trough of 8% in the mid-1970s to its recent peak of 16%...There has been a nearly 10% drop over this same period in the share of national income in the United States..."

I immediately became a COST fan. Should the COST be implemented one day, can we accept its ensuing ripples such as lowering the bubble-like housing prices and commodity prices to avoid the 7% annual ownership taxation resulting in the capital value of everyone's private possessions such as notebook computers, houses, stock shares and savings being reduced by half in return for a regular monthly allowance from the government, i.e. a few tens of thousands of NT dollars per month (insistently derived from the 7% COST paid by the rich). What is our stance in this? That would be another matter.

Radical Markets: Uprooting Capitalism and Democracy for a Just Society
By E. Glen Weyl, Eric A. Posner.,
Chinese Translation by Zhou YiFang
Chinese Edition published by Gusa Publishing on 2020/04/29
Extended reading for Chapter 1 of the book

3 WHEN GREEN CARD LOTTERY MEETS DIPLOMATIC TALENTS

A Hong Kong friend of mine is a wizard of games. He started his own mobile games company 20 years ago. There was no Java Virtual Machine (JVM) back then, let alone iOS or Android. His company made a big fortune after a few years and continuously received investment from several renowned global venture capital firms. He had fantastic fun with his business and was a well-known figure in the industry of games. A while ago, he pulled out and went into a brand new market to begin another venture: Blockchain Likecoin and Liker Land. He ran his new venture spectacularly for a few years attracting over 10,000 users, half of whom were based in Taiwan. Taiwan's users were very fond of his innovative services which had successfully prompted Taiwan's creators to develop many new original content on his platform. Some creators' ecologies had become more streamlined and well circulated thanks to the bloackchain services that he developed. This Hong Kong friend of mine had high hopes over Taiwan's free and mature market that is willing to accept new things. He hoped to have his new company based in Taiwan for long-term development.

In the beginning, this friend successfully joined a famous startup acceleration program in Taiwan; later, he put together some of his previous business investment records to secure an entrepreneur visa in Taiwan. One year later, due to the nature of his company, he could no longer renew his visa to stay in Taiwan. He once asked me: Is there any visa application route via collaboration or recommendation so that he could contribute his strengths while staying here in Taiwan on a more flexible and long-term basis? We had many joint projects, often got together and chatted over anything under the sun. Naturally I would very much like him to stay in Taiwan. In the next few months, I enquired over many official routes and ways for him to stay. He himself made a lot of efforts too. Due to various reasons, he didn't get to stay in the end. This spring, he left Taiwan, a place that he took very seriously and was very fond of, and returned to Hong Kong to carry on with his endeavors. On his next visit to Taiwan, he'd be going through a different immigration gate that would impose more restrictions. It got me thinking that there must be quite a lot of outstanding foreign workers in Taiwan like this friend of mine. If there was a mechanism whereby I could submit to the government a recommendation letter based on my expertise in the industry or in the academia in favor of his stay in Taiwan, I was certainly willing to do so.

I have another friend, who was born and raised in Taiwan and now works in the US. After graduating from the Electrical Engineering Department of National Tsing Hua University, he went on to study in the US and later secured a visa to stay and work. Now he serves in a very successful AI startup. Every now and again, he'd try to persuade me to submit applications to teach in the US, and say that he'd write recommendation letters for me. Besides passing me information about universities and immigration law firms, he would periodically send me pictures of him driving his roadster at 5pm after work to sunbathe on a beach near his company in San Diego. That was what he did throughout the week. As I still have projects that I want to do in Taiwan, I have not got round to those emigration strategies he sent me. One day, he sent a message saying that his sister was migrating to the US. The sister went through a very uniquely peculiar route, not through the O1 or V1 plan but through the lottery. For over 10 years, this friend had been routinely and punctually visiting the online immigration lottery website on behalf of his sister and continued to do so even after he himself had secured a job. His sister got lucky with the lottery, and a few months later, he told me about it. He said with excitement that his sister was soon to move to the US and would be staying with him in San Diego before getting a job.

On hearing this, I was quite surprised. I had long heard of this green card lottery thing since I was a child. I always thought it was an internet scam. Little did I know it was all legit. The friend said that of course it was legit. Its official name is "Diversity Visa (DV) Program", designed to increase diversity in the U.S. immigrant population offering 50,000 visas per year. As this is a very popular immigration route, what with limited availability and vast numbers of applicants, there are numerous online scams trying to con people by claiming it's easier to be awarded such visa with fabricated applications. The lottery program itself is legitimate, but there are many online fraudulent versions of it. Two years later, over a meal with this friend, he sighed and said that his sister had decided to come back to Taiwan for no other reason than that she could not find a job in the US (as her entry was granted by lottery and she did not have a US degree) and that she was not used to the American culture, food and environments.

After a big fight with the brother, she decided to settle in Taiwan and find a job here.

On hearing this story for the first time, my reaction was: what a pity to give up such a rare opportunity of having won the lottery. But then, after much thought, the sister's decision came as no surprise at all. It was not her own intention to draw the lottery, and she only knew one person in the US, i.e. her brother, with all her other family members and friends in Taiwan, so it was only normal not able to fit in. And yet this means this friend's sister had become a case of low-efficiency or zero-efficiency talent mobility. Here we are in 2020, and the United States of America that has been well reputed for its advanced data management and national strategic thinking should still resort to annual green card/visa lottery to balance the immigration population. Is there not a better and more efficient way of balancing domestic or international talent or population migration? That was my question at the time.

In *Radical Markets*, the two authors devoted a lot of space to the exploration of international talents, population and labor migration. In this chapter, they first point out a very important observation and research result, namely, the gap between rich and poor, which is happening not only domestically but also worsening between countries in recent years.

"...persistent differences in mass living standards across countries were unknown until the late nineteenth century. Even the most extreme gaps, such as between China and the United Kingdom, were only a factor of 3. This contrasts with the 10 to 1 gap that opened up by the 1950s."

"Inequality across countries increased from about 7% in 1820 to about 70% in 1980...Together these patterns imply that inequality across countries has gone from a relatively insignificant phenomenon in the grand scheme of global inequality, (accounting for only a little more than 10% of global inequality in the 1820s), to being the dominant source of global inequality, (accounting for two-thirds or more in the second half of the twentieth century and still today accounting for 60–70% depending on whose measurements you rely upon)."

So here is their proposal: if there can be an increased quota of talent mobility in rich countries, and competent workers in poor countries can easily move to and work in a more socially advanced country, this would enable *"...roughly a 20 percent increase in global income"* according to the authors' calculation.

No wonder their views are considered radical in the radical market-ism: what exactly can be done to enhance talent mobility? The two authors propose a radical chapter under the headline, "Uniting the World's Workers" plus an equally radical solution, Visas Between Individuals Program (VIP).

What is radical about the proposal? The authors propose the solution to talent mobility is by uniting the world's workers as is spelt out in the headline. Those with some knowledge of social science would perhaps immediately pick up something odd here. Doesn't the above headline sound somewhat familiar? That's right. A second look would remind one of the Communist slogan that took the world by storm back in the days: "Workers of the world, unite!" The Communism allusion may be missing in the Chinese translation but the English text, "Uniting the World's Workers", is a syntactical rearrangement from the original Communist slogan. No wonder this chapter and its content have attracted much criticism against the two authors. Many readers of the free market faction, on reading this sentence, would inevitably let out a scornful remark, "Communists making a comeback?" Once they read the content, they will probably faint out of shock (or anger).

"Anthony learns of a new program by the State Department that allows him to sponsor a migrant worker. But what's in it for him? Unlike Google, he can't simply place the worker in an office and expect him or her to generate revenue for him.
... the two agree that Bishal will work for Anthony for one year in the United States...Anthony has to use his savings to buy Bishal a flight ticket. They agree that Bishal will reside in Anthony's spare room...If Bishal disappeared, Anthony would also be fined. We don't think the fine should be steep but should be enough to hurt."

Isn't this a renewed version of servitude? Not only will Communism make a comeback but slavery is also to return?

To be honest, this is the chapter which I found most dubious when reading the book. Besides appearing to be highly idealized, it seems to have leaned too much towards data calculation and strategy deployment to have overlooked our accommodation and respect for equality. Reading this chapter elicits the questions: What is happiness? What is an ideal nation? Does an increase of national income equate happiness? For the sake of a narrowly defined happiness, does every worker in not-so-rich countries genuinely desire to work as a migrant worker in Europe, the US and Japan?

"While such an outcome is far from true equality, it is probably the best

that can be hoped for in the near term."

The authors confess near the end of the chapter that they are very clear that the content discussed in this chapter is quite far off the society's current or ideal conception of equality. But the authors' ambition is not merely about exploring the highly skilled talent mobility strategies in various countries in recent years but expanding to examine the low-wage low-rung manual laborers that have already migrated to advanced rich countries. That's why the authors propose their view as necessary despite its temporary shortcomings. They authors think that regardless of whether we go ahead with their proposal, i.e. the VIP, the low-wage labor predicament has already been existent. It's only that the rich class who have a say in such policies in wealthy countries would rather turn a blind eye regarding such matters. The result is that such issues pertaining to labor induction and management are tucked away in a small number of bottomless evil foreign labor rings and routes. These low-wage foreign labor are often bereft of security and end up doing more and more high-risk jobs. Yet there are no appropriate management and combined measures in place to tackle the relevant problems. Instead of leaving the management to remain out of sync, it's better to implement more systematic, mutually-beneficial measures so that those with discernibility can act on such incentives as bonuses or remunerations and bring into their own country workers that are more productive and better protected. This is a stone that kills two birds in the sense that wealthy countries can have influx of better and more talents into all levels of employment, and workers in developing countries will have more opportunities to work around the world as one unified job market in return for an income that they can send back to their home country to contribute to its growth and bridge the income inequality between wealthy and poor countries.

Here are the ultimate questions: 1. Will the VIP attract the intellectuals with discernibility or will it only appeal to the likewise low-wage native workers who are in it for some passive income and indiscriminately bring in incompetent workers? After all,

> *"Our aim is to involve working-class people who would be attracted by the financial benefits of sponsorship. A low-income person who could net $6,000 from sponsoring a low-skilled migrant worker would significantly increase their well-being; in contrast, a middle-class or wealthy person is not likely to find such an opportunity attractive."*

And 2. What incentives are available for wealthy countries to open up borders to even more workers from developing countries than before? Is the VIP system overly idealistic in a social climate dominated by conservatives?

Regarding both questions, the authors first mention the J-1 visa program that is being implemented in the US to prove that the sponsoring system will be popular and effectively managed.

"While the J-1 program was initially designed for cultural exchange, Congress has permitted its use for what is essentially low-wage nanny work... While some people argue that the au pairs are exploited, we have not found any rigorous studies that document abuses."

"They rely on intermediary institutions (private companies) to match American sponsors and foreign workers, train house helps and follow up on workers' employment and housing situations upon their arrival in the US, which is all regulated and managed by the State Department. (Note: This approach is similar to Taiwan's system on migrant workers and house helps.)"

The epilogue in the book carries a supplement that clears my doubt:

"Further, suppose wealthy countries and poor countries reach an agreement where the new VIP system will replace today's practice of using relief fund as a form of assistance to poor countries, and the VIP dictates that all countries agree to share part of the COST income. Given such a premise, wealthy countries can supply relief fund to poor countries at any point in time. However, if poor countries become richer as they develop, such transferred payment will stretch out and the bilateral payment will become equalized. As such, this would give citizens of rich countries an incentive to develop poor countries, as well as giving citizens of poor countries a reason not to resent too much the prosperity of wealthy countries. Together these two features would help tilt the scales of opinion in wealthy countries in favor of opening migration further to aid the development of poor countries."

I could finally understand why it'd be a radical and pragmatic solution if we can replace the Dollar Diplomacy that has been much criticized recently with Talent Diplomacy. If we see education as the foundation of a powerful country, then talent mobility would be the backbone of a country's vibrant economy (the early development in the US refers). One can even say that a further transparent and free talent mobility with reinforced incentives will become the catalyst for a global unified economy and go on to promote a balanced and sustainable economic development for all countries, despite some slightly "ruffled sensibilities". ***"This is a moral gain relative to the hypocrisy of our current system and perhaps the only plausible way toward a more just international order."***

Radical Markets: Uprooting Capitalism and Democracy for a Just Society
By E. Glen Weyl, Eric A. Posner.,
Chinese Translation by Zhou YiFang
Chinese Edition published by Gusa Publishing on 2020/04/29
Extended reading for Chapter 3 of the book

4 HAND GRAPPLING TECHNIQUES OF THE MONOPOLIZING OCTOPUS

I recall that in my youth I used to go to great lengths just to buy computer parts in Taipei's famous Guanghua Digital Plaza; I wanted my self-assembled computer to run faster so that it can download more software for new games. There was no online malls back then. So it was the heyday for the Guanghua Digital Plaza. A short Bade Road was lined with many dozens of 3C shops, with a few more dozens of them in its basement, the "International Electronics Plaza". For the sake of saving a few NT dollars, I'd go through all the shops looking for the best bargain, asking the shop owner, "How much is your 7200 RPM?" or "How much is your 1333 RAM? Is it in stock?" The shop owners were fed up with all these questions and started to put on densely arranged lists of price quotes which were updated daily. When new products were launched, crowds of computer users would swarm outside the shops peering at the lists with much the same emotional eagerness and intensity of students trying to elbow for a view of the result of the university entrance exam. I'd stretch my neck to see if my desired item was on the list, took notes, and compared prices. Sometimes I had to visit over 10 shops just to find a bargain that saved me NT$100, and I'd be over the moon when I did. Sometimes, after a whole afternoon, I only managed to save less than NT$50, and worse still, when I returned to the shop for that cheaper-by-NT$50 bargain, that item might have been scooped by the person right before me. Imagine walking for 10 minutes to get back to the first shop only to be told, "Sorry. Out of stock." That was how I spent my winter and summer holidays enquiring prices, assembling, dissembling and re-assembling the computer.

This went on till one day when I accidentally read a report on a magazine that struck me like thunder, burning me like charcoal, banishing all my remaining youthful innocence. Written in bold type in the magazine were words such as "Many shops in Guanhua Digital Plaza actually belong to the same boss as part of a chain…" Another report stated, "Consumers would previously go to Guanhua Digital Plaza to enquire and compare prices, when actually many stores are under the same ownership, so the prices have always been controlled by the sellers. Such asymmetric information has earned immense profits for some business proprietors…" So when I was zipping in and out of shops burdened by my sachet of school books hunting for the best deal, those "best shops" with "honest pricing" actually did not exist. Those shops very likely belong to the same person or group of persons. These people open up a bunch of shops to sell identical goods. Since customers like to compare prices, out of necessity and out of habit. "Love comparing prices, huh? OK you can have a field day. Some shops would be NT$10 cheaper or pricier. No matter which shop you go to, you'd still fork out the money and it all goes into my pocket." Such practice has been around till the emergence of online shopping.

Here is another example. I believe many readers still remember the panic buying of toilet paper a while back. As recorded in Wikipedia: "Taiwan's panic buying of toilet paper in 2018" dubbed as "Toilet Paper Chaos" or "Defecation Chaos" refers to the phenomenon of panic buying of toilet paper following speculation of price increases caused by manufacturers' indecent marketing techniques and media coverage. Back then, the Fair Trade Commission did eventually launch an investigation on toilet paper manufacturers to determine if their joint act of price hikes was illegal. There were also the 2011 incidence of Taiwan's top 3 dairy companies having a joint price hike on fresh milk (the result being that the top 3 companies owning a combined 80% of Taiwan's dairy market were fined NT$30 million in total by the government) and the continuous joint price fixing by CPC and FPC for 22 times between 2002 and 2004 (with both companies being fined NT$6.5 million each).

Some of the above cases are about shops having the same owners while some others are about various entities hooking up as market makers. These enterprises as daily services/goods providers can often be seen joining forces to increase their profits, and the general public can only respond by comparing prices or falling into shopping frenzy. But what if these companies on a monopoly strategy have long infiltrated deep into the bottom of the social structure and capitalist services?

There have long been conspiracy theories spreading the claim that the

American society and capitalist structure are controlled by a small number of families or enterprises. Today, Chapter 3 in *Radical Markets*, finally unveils some of the truths about this rumor. The top six banks that control all the monetary services in the US share many of their top 5 shareholders. These banks that appear different for carrying different brands and strategies and even competing against each other for customers are yet jointly owned by a smattering of holding companies. Such revelation from seeing the chart in the book for the first time hit me like a bolt of lightning akin to the memory of wasting my youthful days in Guanggua Digital Plaza and being conned out of money by the same bosses.

"Since the late 1980s, BlackRock, Fidelity, Vanguard, and State Street have not just grown large in absolute terms. They have also become the largest shareholders of major US corporates.
The top 5 shareholders of the six largest US banks:

JP Morgan Chase / BlackRock, Vanguard, State Street, Fidelity, Wellington

Bank of America / Berkshire Hathaway, BlackRock, Vanguard, State Street, Fidelity

Citigroup / BlackRock, Vanguard, State Street, Fidelity, Capital World Investors

Wells Fargo / Berkshire Hathaway, BlackRock, Vanguard, State Street, Fidelity

U.S. Bank / BlackRock, Vanguard, Fidelity, State Street

PNC Bank / Wellington, BlackRock, Vanguard, Fidelity, Barrow Hanley"

*See if you can spot the name of a shareholder that appears only once.

Many readers may recall the term, "antitrust law" which seemed to have protected the world aiming to dismember the octopus (the monopoly) that single-handedly dominates the service market. Indeed, AT&T and Microsoft's IE browser have all been investigated under the antitrust law for monopoly. The antitrust law achieved great success in 1984 in that AT&T was dissembled into an offspring of the parent company, namely, a new AT&T, plus 7 local telephone companies. In the late-1990s Microsoft case, we can see that the antitrust law enforcement came across a setback: in 1998, the US Department of Justice filed a charge against Microsoft for compulsively embedding the IE browser in its operation system which had Microsoft entangled in a lawsuit for antitrust law violation for many years. A settlement was reached in 2001, which was to some degree a triumph on the part of Microsoft.

In recent years, the US Department of Justice has been targeting Facebook, Google, Apple and Amazon with stringent antitrust law enforcement, particularly in 2019. Look at the recent large merging cases, e.g. AT&T merging with Time Warner; Disney merging with Twenty-First Century Fox and Marvel Studios. They are all mega merging projects leaving us exclaiming in awe. In the past, we could hardly imagine that these characters like Woody, Elsa, Darth Vader, Iron Man, X-Men, Avatar, and the Simpson would go under the same company as the Mickey Mouse. The overall value of the intellectual property (IP) of these fictional characters is said to be on a par with the wealth of a nation. The US Department of Justice was at some point concerned about Disney's merging projects, but eventually filed no charges. We can see that the worldwide trend has been consistent with the conclusion proposed by *Radical Markets* authors: the anttrust law, with its good intentions, was once formidably functional but is now no longer a force to reckon with.

"Because of this activism, American antitrust law became a model internationally: it spread first to Britain and then to the European continent and farther around the world. Yet just as American authorities gained the admiration of the world, they stepped off the Red Queen's treadmill. Beginning in the 1970s and accelerating from the 1980s onward, antitrust authorities lost track of the ways in which capital markets reconfigured themselves to maintain monopoly power. In order to understand the reasons why, we must examine the evolution of the corporate form and its governance in the United States during the twentieth century."

Here we are today in 2020; Facebook, WhatsApp and Instagram have close to 3 billion users in total, and all three companies are governed within the same decision-making system. Put simply, Facebook alone has control over the top 3 out of the world's 10 largest social media platforms. In what appears to be a competitive industry, there is essentially little or no competition. These mega enterprises once again act like an octopus gripping the economy and the society, monopolizing any market they want to tap into. This has solidified the power of those possessing contemporary capital and will drastically diminish consumers' right to speak up. The inexplicable disappearance of data as power as discussed in Chapter 5 is also to do with monopoly as illustrated in this chapter.

"Walrus saw private monopolies (along with private land

ownership) as both the primary impediment to the operation of free markets and the central cause of inequality, writing in the 1890s, 'Look in American for the sources of the enormous fortunes of multimillionaires... and you will find...the operation of business without competition.'"

Do you want to live in this kind of world? Do you want to end up in a world where everything is owned by the same bosses like all the shops owned by the same people in the Guanghua Digital Plaza? Do you hope to endlessly compare prices just to save more money and to consume more efficiently, when all the prices are jointly fixed under the table by sellers to exploit the rights that you are entitled to as a consumer? In an era of global management and digital communication, we can foresee that the service platforms for industries such as entertainment, finance, science and technology, airlines, automobiles, transportation and accommodation will further expand (e.g. Uber and Airbnb) to such an extent that the all-gripping, all-linking octopus can more easily hide behind all the massive information overload.

One of the authors, Eric A. Posner, is the eldest son in a family of legal scholars (Eric's father being the most-cited legal scholar of all time based on The Journal of Legal Studies). Posner is tremendously intelligent with outstanding employment history. After much deliberation, he proposes in the book the idea of resuscitating antitrust laws and tries to remind us via the term, "Monopsony" that monopoly does not appear only in Wall Streets or indeed on any streets but also in the job markets. Have you ever thought about why wages for the middle class are slow to rise, while the upper-income class often enjoy a quantum leap in their salaries? The above is actually an antitrust act. Those enterprises that pay for our labor can effortlessly achieve their goal of saving labor cost simply by passing joint austerity policies. And why would they not, if they are not being confined by compulsory law enforcement.

Radical Markets: Uprooting Capitalism and Democracy for a Just Society
By E. Glen Weyl, Eric A. Posner.,
Chinese Translation by Zhou YiFang
Chinese Edition published by Gusa Publishing on 2020/04/29
Extended reading for Chapter 4 of the book

5 EPILOGUE: EXTENDED READING WITH LOCAL PERSPECTIVES

After reading this book, I was unsettled beyond words.

Numerous words flashed through my head. If there is to be a new ism, what word can best describe it?

Infosocialism?
Telecommumarketism?
Socialmarketism with free-flowing information?
Or Liquid Democracy

In Chapter 1, I saw resources (assets) liquidity.
In Chapter 2, I saw democracy (votes) liquidity.
In Chapter 3, I saw talent liquidity.
In Chapter 4, I saw power liquidity.
In Chapter 5, I saw data liquidity.

The more liquidity there is, the more freedom and equality there will be in the world which will tip towards the ideal prospect as proposed in yesteryears by those holding free market-oriented views: the market will automatically drive towards an overall optimization and equal allocation.

If the market is to become what the authors shockingly predict it to be in the epilogue: the market will perform as one computer, and liquidity as algorithm, then we can imagine that today's market will morph into a gigantic motherboard or IC chip of a massive volume and size, owing to the world getting flatter with hyperconnectivity. But the liquidity volume of market information along with its connectors (persons and entities that hold such knowledge or information) is somehow diminishing, compared with the past. The market information volume and complexity is on an exponential growth, while people's information handling and comprehension ability is on a linear growth. This will result in a peculiar phenomenon where the minority complies with the majority, and the majority is controlled by an even smaller minority.

What can be done?

Build a computer that belongs to everyone. The computer here holds two connotations: one being an actual computer, through which humans will for the first time have control of a documenting system and fair smart collective decision-making strategies and it will be sufficiently trustworthy, transparent despite its currently insufficient efficiency. The other connotation is about building a market that genuinely belongs to everyone, instead of having a small number of people singularly controlling or even monopolizing the flow of information and decision making

In the end of the book, the authors let their imagination fly and plan a nearly all-knowing almighty (and radical too) computer. Given their economic calculation, they hold the view that with such a huge computer, there will be justice hereafter in the human society, or phrased differently, equality in the world. Every manufacturer anywhere in the world will be informed at all times the demand status of those in need on the other end of the world and can better adapt to the weather conditions in source regions somewhere else so as to timely adjust manufacturing quantity, supply and prices. Viewed from another angle, this computer is no different from an AI mastermind in sci-fi fiction.

Is such a future good or bad? Do you want to trust a super AI computer that anyone can log on to debug at any time? Or do you chooe to believe in the country leaders elected by each country via Populism-influenced voting in our time? I have no answers. But I do believe that as the world carries on evolving as it is today, we will surely develop the aforementioned computer, which however may not belong to us the people, and the so-called market will eventually devolve into a mirage for an indefinite period of time.

6 ONE THOUSAND WAYS OF THINKING FOR THE 21ST CENTURY:
READING *21 LESSONS FOR THE 21ST CENTURY*, Y. N. HARARI, AUTHOR OF *SAPIENS* AND *HOMO DEUS*

Originally published on Medium

I often think about this: what will happen if this world is without thinkers? Ayn Rand, an American writer and philosopher, turned such musing into a three-part philosophical novel, *Atlas Shrugged*, depicting an anti-Utopia parallel universe in which thinkers and creators all disappear or go on strike because of deliberate human manipulation and the social systems. If thinkers all over the world go on strike, because of governmental control and looting, and they decide to stop inventing new things, creating art, conducting business and researches, what will happen?

I also often muse over this: what will happen if this world is devoid of futurists? When I was in Tam Kang University, my alma mater included a new course into the general education curriculum for the first time, called Future Studies. When selecting courses and seeing this new addition, my fellow students and I were all looking at each other, puzzled at what this course might contain. We even jeered at those who went for this course as Sci-Fi Nerds or derailing off the academic path. It was not until 2014 when I was fortunate enough to enter Singularity University where I met my teachers: the contemporary futurist, Ray Kurzweil, and actionist, Peter Diamandis, from whom I first heard of block chain, smart mechanics and deep-learning artificial intelligence, as well as the hidden impact these technologies will bring to the human society, such as rocketing unemployment rates, financial restructuring, machine killing and so on. If we only know how to steep ourselves in theoretical studies of technologies and manufacture one after another iPhone, Pepper and unmanned vehicles, can we make this world a better place?

21 Lessons for the 21st century, a new book by Y. N. Harari, author of Sapiens and Homo Deus

The book, *21 Lessons for the 21st Century*, has answers for the two aforementioned questions. Both thinkers and futurists are indispensable. The author Y. N. Harari played both roles in the book. First and foremost, he is an outstanding historian. The Apple Inc founder, Steve Jobs, once said, "You can't connect the dots looking forward; you can only connect them looking backwards." On questions about the future, Harari blends his detailed reasoning with history-based discourse on the human society. Layer by layer, he unravels subjects in our full view, tracing back to Nazi Germany, ancient Rome, and even to the Biblical times. He proposes unprecedented, innovative and thorough thinking in response to all major big questions: from artificial intelligence to big data; from nationalism to future religions; from immigration to post-truth eras; from terrorism to sci-fi fiction. He even devotes the entire last chapter to exploring the "Meaning of Life"

These keywords may sound a bit serious and dull, but once you start reading the book, you cannot stop. It contains real life stories that happened around the world or reports that you read on Facebook only a few months ago which underwent the author's excellent deducing and reasoning, and you will see that our brain that has been slowly drying up from non-stop round-the-clock bombardment of FB stories, Crazy Friday, will be plowed and stirred open; all the undeveloped cognitive cells will be turned over and injected with life spring.

On Universal Basic Income (an emerging social system advocating that people should be entitled to an income to live on owing to vibrant mechanic automation that allows people not to work), he not only explores the existing planning of systems but also reviews the definition of "universal". Right. What is universal? At what age do you qualify as universal? Will those who have gone to Mars count as universal? What about robots? On "fake news", he discusses "post-truth" in many chapters. Besides the latest news on the privacy invasion concerning Google in Europe and Facebook in the US election, Harari also includes religious and historic thoughts in his discourse that the prosperity of religion is derived from people's pursuit of and belief in post truth. Absolutely! We may believe in gods, but those who don't might possibly see biblical stories as fake news! After all, who has seen a deity? What is the truth? What is faith?

Harari says, "Truth is often a burden, while rituals and ceremonies are somehow your best buddy." I instantly got it; the centennial Confucius thoughts perhaps bear negative effects after all. On automated vehicles, he devotes much space to the discussion of trolley problems, a topic that is seldom talked about in Taiwan. Suppose an automated car cannot avoid clashing onto two children that charge into the track of the car, and inevitably only by sacrificing the car driver can the children be prevented from injuries or death. How will the decision making be conducted by the algorithm in this instance?

Ethical scenario

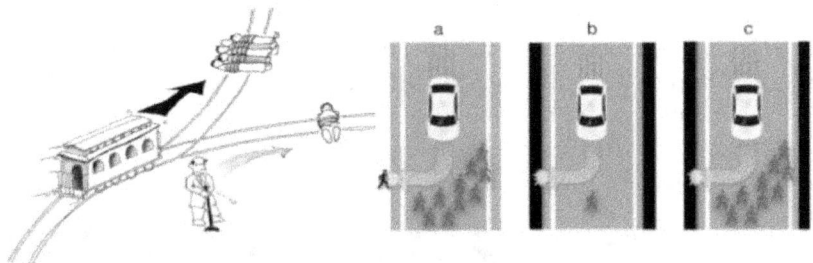

1. https://medium.com/@tanayj/self-driving-cars-and-the-trolley-problem
2. http://www.businessinsider.sg/the-ethical-questions-facing-self-driving-cars-2015-10

trolley problem

On money, Harari and I shared the same view, given what I have learned about block chain from the past two years.

"At first glance, religious texts and money appear to be two separate things, but in fact they are very similar conceptually. Most people, on seeing a US dollar bill, will not remember it is only part of a system agreed upon by humans... They find the bill itself valuable, rather than reminding themselves that 'This is a useless piece of paper. Only because other people think it is valuable, I can now make use of it.'"

Such thought blew my mind. The moment when I saw through all this was as astonishing as the moment when Neo suddenly wakes up and unplugs from the mushy incubation pod in the movie, Matrix.

There has recently been some sudden appearances of low-brow red-yellow-blue clay cartoon videos on YouTube tricking children for clicks. This is commonly referred to as Elsagate. There were already conspiracy theories calling out Disney for orchestrating the Elsagate. Harari used capital letters to illustrate his sufficient evidence that Disney no longer has faith in free will. This is astonishing! Seeing that Disney is acquiring Star Wars, American hero movies, Fox Entertainment as casually as buying daily groceries and that almost all my idols and IP with a cultural and AV impact are going under the same production company as that of Mickey Mouse , I cannot help but get as concerned as the author and wonder what the future holds for us.

To quote Kevin Kelly, the editor of Wired Magazine, for his view on "Technology":"Technology is like an organism. At most you can influence it but cannot stop it from advancing. We call history a giant wheel, and we certainly cannot stop this giant wheel under the giant bandwagon built with a combination of key components like science, history and society. But after all it is humans that are steering history that is the bandwagon, i.e. societies constituted by humans, super egos and superheroes co-created by multiple social consciousness. Some call it divinity. I rather call it humankind."

"A vintage dial and spin telephone" by <u>Daria Nepriakhina</u> on <u>Unsplash</u>

History is not linear but an old-fashioned coiled cord, forever heaving and rolling, appearing to be full of sunshine in one minute and falling off the valley in the next. But we must believe that as long as we do not stop thinking, much like what Harari says, "The future workplace will be in further need of philosophers." As long as thinkers don't go on strike or are made to disappear, and the creators continue to have serious and responsible discussions, while gazing towards an alternative distant future and exploring multiple possibilities rather than fixing on one answer, then this winding cord will one day lead us to that handset and whatever is behind the handset, so that we get a chance to hear whether "the answer to the ultimate question of life, the universe, and everything" is 42.

7 FOLLOW COLUMBUS IN THE MOMENTS OF HISTORY - A FOREWORD FOR TIM O'REILLY'S NEW BOOK *WTF? WHAT'S THE FUTURE AND WHY IT'S UP TO US*

原文發表於《未來地圖》臺灣中文版

A Legend of Time

Tim O'Reilly is the founder of O'Reilly Media, unique among tech book publishers for its iconic animal covers. Tim is a strong advocate of free software and the open source software movement. He invoked the term, "Web 2.0," and has been propagating the forward-looking notions such as "Gov 2.0." Tim is my idol. I am very grateful to CommonWealth Magazine for inviting me to write a foreword for the Chinese translation of Tim's new book *WTF? What's the Future and Why It's Up* together with **Kai-Fu Lee (Chairman of Sinovation Ventures)** and **Jung-Ting Yeh (Chairman of FamilyMart).** At the end of the article, a discount code is specially offered to my readers for use on Eslite online and Pok'elai. Don't miss it!

Meeting Rooms Are Think Tanks

I often share with students how to train themselves to predict the future and guide them through the way how great creators think. The method is very simple, and I call it "thinking in a meeting room." Imagine a conference room full of people who made history. In this conference room, they brainstorm for hours and discuss how to create something that has been created today.

Employees of Pixar meet to have an internal discussion on Toy Story. (*pixar.wikia.com*)

Through this method, you act like a retrospective detective or disassembly engineer who goes back to and guess at the historical result. Because you already know the result, it is possible for you to compare the result and infer the possible process and important decisions made in that meeting. Although it is a little bit difficult, it is definitely easier for you than for people who are actually in that meeting room because they don't know the result!

I call the meeting room having popped up in my mind many times that gave birth to iPod, iPhone, the first 3D animation, Facebook, Amazon, and the movie, The Fifth Element "that meeting room." The more stories, history, and details in "that meeting room," the clearer we look into the future. We can even create and enter such a meeting room to make history.

The book intrigues me right from the first chapter. Every paragraph is like "that meeting room" constantly reappearing and reorganizing.

Origin

When the term, "Open Source," did not come into existence, the author explained how he had discussions with a group of pioneers in the technology industry and created such a huge historical movement that has shown no signs of slowing down to this day.

The author also explained how he and his friends gave birth to Global

Network Navigator (GNN), the first content-based website acquired by a company in the Internet era, in the early 1990s when Yahoo was still in its infancy or precisely dated a private e-mail to Jeff Bezos, the founder of Amazon at **10:03:59 AM on Wednesday, January 5, 2000.** This e-mail eventually gave birth to BountyQuest, further inspiring Kickstarter and other crowdfunding platforms around the world.

Steve Jobs and the Mac team have a lunch meeting. (*cultofmac.com*)

When I read this book, the secret moments that have such a huge impact on our modern lives are unveiled. **That/those meeting room(s)** must exist. We just have to delve deep in order to enter. All the great things, as they are, have a beginning. This book is full of awe-inspiring moments when you find out that something just looked differently at that time, and the author played his part in those moments!

Moments of History

After I finish reading the book of nearly 500 pages, it seems that the author guides me to a place between the present and the future. A miraculous office building nestles in the time corridor, where celebrities from various countries appear to work from "that meeting room" to a full floor office space.

When walking down the time corridor, sometimes you will encounter celebrities such as a ministerial official from Germany in the G20 meeting criticizing Uber, legendary investor George Soros talking to the author about the "post-truth era," Washington Post columnist John Farrell debating on fake news, the CEO of General Electric and the White House administration expressing their views on the next economy to the author at the NEXT global economy conference. In a certain chapter, we can find an example from Taiwan, vTaiwan, led by former minister of state Yu-Ling Tsai, Audrey Tang, and g0v (see g0v.news), which is the best case for the author to explain how **technology is changing politics**.

vTaiwan is a virtual policy exchange platform and a model for digital governance in Taiwan. (*vtaiwan.tw*)

In the last chapter, the author talks about his views on the future. In general, he is optimistic about the future, but he also expresses his concern about the current situation. In addition to concerns and discussions about **"super currencies,"** he also explains and encourages **"universal basic income"** (see the article by Jia-An Chu, a writer of philosophical thinking); in addition, he also points out some unique ideas. For example, he proposes that "future currencies should be divided into two types: **"machine currency"** and **"human currency."**

Reflection

Before war, soldiers carefully read the map lying on the table. Likewise, the author details the frontier technologies over time, trying to lead us to advance and adventure. This map constantly reminds me of the blockchain industry that I have been engaged in recently. Blockchain technology was proposed a decade ago, but there are still many terms that are undefined (e.g., DApp, and Backup Phrase). Comparing to the trajectory of technological developments described in the book, it is like a recurrence of the author's participation in the creation and development of the Internet. This makes me feel more confident of developing blockchain technology, and many scenes and experiences described in the book become the best cross-temporal comparison and reference for my current work.

Conclusion

What our future will look like depends on how we take on the predictions to shape this century. In the past, we might be hesitant and kept in the dark. After reading this book, we will all become Columbus in the moments of history, holding a map full of history, experience, stories and courage that guides us to **a more human-centered future** depicted by the author.

《未來地圖：對工作、商業、經濟全新樣貌，　正確的理解與該有的行動》,Tim O'Reilly, 天下雜誌, 譯者： 黃庭敏, 2018/12/26 Published.

Keenly Waiting Online!
Observation of Contemporary Interactive Culture and Technology-Based Society

8 "CAN CROWDS BE CONTROLLED?"
12 STEPS: FACE OFF WITH YOUR OWN BRAN AND YOUR FUTURE SELF,

原文發表於《12 步：跟自己的腦對決，也是跟未來的自己對決》臺灣中文版

Foreword

I will always remember the MOS BURGER outside Exit 2 of MRT Taipower Building Station. Over there I first learned about the vulnerability of human nature and the precariousness and untrustworthiness of social media. One evening, a few years back, I was in the middle of starting up a business, creating Taiwan's very first Location Based Service (LBS), a mobile social group application called Linkwish. For a few days something had been bugging me. Back then, Linkwish only just went online less than 6 months and had garnered close to 10,000 registered members, fast becoming a popular social media platform. Something big and earth-shattering seemed to be coming my way.

Linkwish is no longer functioning now. So that earth-shattering event, predictably, was not about the flash increase of the user number, which was on the contrary fast plummeting due to some phenomenon. Stuck and confused, I bumped into J, a sociologist with whom I had previously done some research with in the Academia Sinica. On a whim, I called out to him and offered to treat him to a big meal at the MOS BURGER, so I could probe and elicit some insight from him. I got into action and duped J into the burger store where I updated him on my mobile application platform: not much interaction among members, and malicious private messages being sent by some users that were abusing the platform to harass well-behaved users who were then fed up and decided to leave the platform never to come back.

I asked J: Can this be resolved?

J smiled and asked me: What do you think? What'd you say? Do you think "Crowds can be controlled?"

Unprepared, I was taken aback by this question. I could not help but wonder if the answer is yes, then J should tell me how to control crowds pronto. If the answer is no, then it means not only is it impossible to resolve my problem but it is also not worth mentioning. It was a time when there was no such job title as Social Media Manager nor such concept (circa 2010). Musing over it, I said to J as a reply, "It should be possible to control crowds, or else there is no point in asking this question. Right?"

As a sociologist with the most extraordinary mind in the Academia Sinica, J picked up the just served piping hot rice burger in a mysterious manner, took a bite and said, "There are two factions of sociologists; one thinks that not only human individuals but also crowds and societies can all be controlled, while the other thinks that they are beyond control. That is, the latter are optimistic over humans' free will and rational thinking (i.e. humans are unwilling to be controlled and can remain alert at all times regarding if they are being controlled)." Gulping down the iced black tea and fries while gesticulating, he said, "But today you've got the right person here. I happened to be of the former division. I think people or crowds can be completely controlled! As long as you do it the right way, you will obtain the key to controlling crowd behaviors and getting them to do whatever you want them to do."

I was blown away by J's words. If those words come from some kind of internet content farms or the likes of ET.Lemons Facebook page, I might have been able to remain calm. But no, this is a highly acclaimed Academia Sinica sociologist saying "people can be controlled". Though it went along with the proposition of my question, still I was rather startled. Next, J taught me all about social manipulation techniques such as "Crisscross Seating Pattern Control Method", "Opaque Room Method", and "Socialization Prowess Coding Method". But that would be another story.

The book, *12 Steps: Face off with your own bran and your future self*, was written by Dr. Jeong Jae-Seung, a biology and brain science professor at the Korea Advanced Institute of Science and Technology (KAIST) who was once a psychiatry research fellow at the School of Medicine, Yale University, devoted to brain study and decision neuroscience. The book has repeated attempts at exploring the wide-ranging states of the human brain when its autonomous behaviors are under external influences without being aware of them. In the past, such topics would have only involved criminal psychology, medicine, military or warfare. However today humans are suddenly being propelled by the so-called digital technologies and internet freaks onto a singularity high-speed bullet train which is, on a mass scale, at full throttle, charging towards these issues. The entirety of the human social structure and individual behaviors will be instantly impacted by human cognition and behaviors whose causes and details have not been deciphered and illuminated by scientific studies until today.

We are not afraid of AI invasion on humans, not even scared of wars erupting between countries that don't see eye to eye with each other. Now we are more afraid of having our societies and crowds being manipulated to the point of invasion by the AI without being aware of it. We even worry about countries utilizing AI, big data and micro data analysis & control to build one after another shapeless omnipresent matric just like in the movie, Matrix, with a difference that this very first co-created matrix, since the genesis of humans, interwoven with AI, exponentially accelerating technology & neuroscience and behavioral science (including behavioral economics), will not be equipped with a burr hole and that physical wake-up call will be non-existent, and Zion, the symbol of redemption, will no longer be around.

In future, how are we to maintain our autonomy and competitiveness when artificial intelligence is geared to rule our lives, rule our jobs and rule our times? The answer is in the book, *12 Steps: Face off with your own bran and your future self*. Without confronting your own brain, you cannot be clear of what dopamine, endocrine hormones and cranial nerve reflexes are impacting on your behaviors, and how can you consciously and rationally confront yourself, let alone confronting the AI that has long merged with science. Cambridge Gate is the first strike by those with invested interest taking advantage of AI and digital networks upon grasping weaknesses in human behaviors. Should there be another strike, it may not be so easy to detect or fend off. To confront a device like that, the entire human society will have to confront the human brain first. Knowing yourself may not lead to complete triumph in all battles but will at least allow us to speculate what kind of artificial intelligence, digital totalitarian system and digital marketing tools will be used in the next few decades to defeat our human brain by tackling its weaknesses. Having read this book will surely make it easier to defend our brain. We absolutely have to safeguard our "free will and autonomy" as gems that we humans most value and can be most optimistic and proud of.

《大腦革命的12步：AI時代，你的對手不是人工智慧，而是你自己的腦》,정재승,八旗文化,譯者： 謝宜倫, 2020/01/02 Published.

Keenly Waiting Online!
Observation of Contemporary Interactive Culture and Technology-Based Society

9 WELCOME TO ZION; THE FORT OF FINAL TRUTH REQUIRES OUR PROTECTION. PREFACE FOR DEEP FAKES AND THE INFOCALYPSE: WHAT YOU URGENTLY NEED TO KNOW

原文發表於《深度造假：比真實還真的 AI 合成技術，如何奪走人類的判斷力，釀成資訊末日危機？》臺灣中文版

On an evening in the early months of 2018, I came upon a keyword on my social media: DeepFake. Very soon an onslaught of DeepFake-related information swept across internet, most of which was in English without any Chinese translation at the time. Some were Twitter messages inserted with an animated GIF or short video marked with #DeepFake. Its content was mostly well-known Hollywood actresses engaged in titillating movements, e.g. undressing in front of the camera or laughing playfully with adult toys in bed. At first I thought some female movie stars got their phones hacked. Before long, it was covered in the international media that an anonymous computer wizard had recently developed a mobile phone application where it took only a few photographs, a computer display card, a desktop computer, and a few hours before a well-made simulation video of real people was produced. Shockingly, this application was made available recently. This means anyone can download the application for further modification and run it on one's own computer to make any desired video.

Suddenly the entire internet was swamped by rampant circulation of DeepFake videos, examples including Nicholas Cage appearing as Luke

Skywalker in Star Wars, Hong Kong porn stars appearing in Aquaman, to name but a few. The majority of such videos feature beautiful actresses in some private sexual acts; the sheer number of it was very confusing and leaving us uncertain about what is the real news, what are being hacked and what are artificially fabricated by the computer software. Each DeepFake video looks so authentic. As far as forgery goes, it appears so much more advanced than the Hollywood synthesis technology and more integrated than the Ironman technology that even the top-notch expert cannot spot the difference. Within days, the biggest social media platforms, Facebook and Twitter, had no other resorts but to team up and drastically reduce the appearance, exposure and circulation of #DeepFake articles with nearly compulsory measures. As a result, the entire furore appeared to have quietened down. At the time I was focusing on this simulation technology so I immediately downloaded the DeepFake software, researched with my students and came up with two videos, Doctor Bao in Ironman and Doctor Bao as the lead in House of Cards, which have gone from garnering several thousands of likes in the beginning to near oblivion and lack of interest instantaneously. This goes to show that not only Facebook and Twitter but YouTube has also joined the line of blocking this DeepFake technology.

After that, a new batch of DeepFake videos would crop up intermittently. For instance, because of Mark Zuckerberg's refusal to take down DeepFake videos on Facebook, his opponents made a malicious video of Mark Zuckerberg appearing in a Twitter video talking about Facebook violating human rights and personal privacy. After videos of crazy AI attacks on humans spooking out millions of viewers, it was revealed that such videos were also works of DeepFake simulation. About 4 o'clock in the afternoon on January 3, one week before Taiwan's presidential election, a slapstick video was suddenly circulating on internet of a presidential candidate joking and laughing with a Hong Kong movie star, Stephen Chiau Sing Chi. In the video, the presidential candidate commented on the underground betting supposedly related to him and said lots of things unfavorable to his election. Although the video maker deliberately cast an actor that was starkly different from the real candidate in physique, still they were so alike in facial expressions and all that was being synthesized that the two were almost identical. On watching the video on that afternoon, I felt as if I was being hit by a bolt of lightning. I immediately became aware of this possibility: the technology that made this video was so mature that if there were to be any follow-up series, if we leave aside the nonsensical slapstick of mimicking Stephen, and if more actors of similar builds were elicited to play more presidential candidates and say something as shocking as "Both sides of the Taiwan Strait will go to war (or be unified) with immediate effect, it is not difficult to imagine that such a video would send a huge shock wave to or play a huge impact on the election result, or even overturning it could have been possible. I immediately issued an alert across the internet with thousands of people sharing my article of disclosure. One of them was the internet celebrity, Ben Jai. Fortunately the video was intercepted from further circulation and subsequently caused many hundreds of internet users to leave messages on the video demanding an explanation from the production team what the ending caption on the video, "More series to follow. Watch the space" entails. Eventually nothing came out of it.

The aforementioned was only a little instance that happened previously in Taiwan's internet circle. Yet, this book covers a wide range of DeepFake cases that have been circulating across the world, ranging from the most classic, fabricated Putin-and-Trump videos to the latest and first DeepFake commercial in April 2020 that forged content and voices from an old video footage in the 1990s to show a news anchor broadcasting news 30 years ago and predicting the future in 30 years' time. Netflix has launched the truth behind NBA documentaries, not to mention the technical aspects of the DeepFake that has kicked up a wave of fake news, fake truths and post truths as initiated by political figures. The book also covers a great detail of the black man, George Floyd incidence that erupted this May. It also covers the news of the US president telling the media that injecting disinfectant can work as a treatment when the world has fallen into the worst crisis of the century due to the virus outbreak since early this year. It notes that not only visual deception, news setting but also one's voice on the phone can all become the DeepFake battlefield. In the 2018 annual Google convention, a technology named "Duplex" was launched. It was demonstrated on the spot that AI-synthesized human voices could make a reservation with salons and restaurants on the phone. When the two tasks were completed, the entire auditorium erupted into a thundering applause. To everyone's dismay (or surprise), it is never easier for AI to mimic humans speaking.

Since 2018 I have been following fake news, fake videos, fake voices and all kinds of forgery techniques. In addition to having a go at the synthesis technique myself, I have also been writing articles to raise more concern and more discussion. After all, the era where "seeing is believing" will soon be a thing of the past, and yet many people still remain in the old era and maintain the old perception, totally unaware that in this digital era, the computer technology, if manipulated by the wrong people, can create almost any people or objects that you wish to see. It is dead easy to forge a teleconference call using the existing technology and show you a sight of a family member or a friend being wounded, tied up and crying for help at the other end of the call. Likewise, it would be easy-peasy to show you a political figure that you support appearing in a communist country and submitting a letter of surrender and declaring subjugation to the communists. It has become so much easier to show you all aspects of "truth" in any way possible. Now what measures can we take to verify truth? At the end of this August (in 2020), the crazy entrepreneur, Elon Musk, launched the first-generation brain-computer, Neuralink, claiming that in future music can be transmitted directly to the brain and that language transmission no longer requires opening your mouth. Foreseeably, all future color images across the world can be shown to you within that little device. Even palatal and tactile sensations can be absorbed by the brain via that tiny device. That is to say, in future seeing is no longer believing, and even objects cannot be authenticated merely from the tactile contact. So how are we going to cope in a world where everything and our experience with the world can all be forged by DeepFake and transmitted directly to the brain?

This book marks the start of moving towards a new world. Despite a painful read, the book will enable us to maintain our truth and insight. The way to sift through fake information is to first keep informed of the current forgery technology development. As stated in the last chapter of the book, to counter a future of DeepFakes is to begin with the first step, "Understand", followed by "Counter" and "Combat". Come on! The dark future when simulation technology gets out of hand has quietly arrived. It's time that our end-game against the fake information started. Let us alert each other and remain awake in defending this fort of truth that I call Zion. It demands that we all step up and safeguard the truth.

《深度造假：比真實還真的 AI 合成技術，如何奪走人類的判斷力，釀成資訊末日危機？》, Nina Schick, 拾青文化, 譯者：林曉欽, 2020/10/14 Published.

ABOUT THE AUTHOR

Ju Chun Ko now works as Assistant Professor in Department of Interaction Design, National Taipei University of Technology.

In 2012, Ko graduated from the National Taiwan University with a Ph.D. degree in Computer Science. In 2014, he received National Science Council subsidiaries for his postdoctoral research in the Keio Media Design (KMD) graduate school in Japan's Keio University. Later he became the first Taiwanese to study in Singularity University, said to be the "smartest university in the world." In 2015, he co-launched the Luna 360VR Camera campaign on the crowd-funding platform, Indiegogo, raised US$350,000, received investment from HTC and won the Red Dot Design Award (of the world's top 3 design awards) in 2017. After that, he started to get actively involved in the promulgation of the blockchain industry and its relevant schemes, e.g. the smart contract development initiative, while assisting many corporates exploring blockchain crowd-funding, digital assets transaction, blockchain games and other projects.

www.ingramcontent.com/pod-product-compliance
Lightning Source LLC
Chambersburg PA
CBHW070857220526
45466CB00005B/2023